A World of Plants

Kate Boehm Jerome

PICTURE CREDITS
Cover (front), Bill Hatcher/National Geographic Image Collection; 1, 13 (bottom), 18-19 (middle), 19 (right), 26 (top), 35 (bottom), The Image Bank/Getty Images; 2-3, Richard Nowitz/National Geographic Image Collection; 4, 4-5 (top), 27, Botanica/Getty Images; 5 (top left), 34 (second from top), Owen Franken/Corbis; 5 (bottom left), 34 (top), Medford Taylor/National Geographic Image Collection; 5 (right), Greg Probst/Panoramic Images/NGSImages.com; 6-7 (background), 7 (second from left), 34 (second from bottom); Stone/Getty Images; 7 (left), Richard Herrmann/Visuals Unlimited; 7 (second from right), Ed George/National Geographic Image Collection; 7 (right), 31 (top left), 31 (bottom left), 34 (bottom), Royalty-Free/Corbis; 10 (top), Phil Schermeister/National Geographic Image Collection; 10 (bottom), 35 (second from bottom), Robert Pickett/Corbis; 11 (top), 25 (top right), Anne Keiser/National Geographic Image Collection; 11 (bottom), 31 (bottom right), Photodisc Green/Getty Images; 14 (top), David Muench/Corbis; 14 (middle), ML Sinibaldi/Corbis; 14 (bottom), 15 (right), Jeff Rotman/jeffrotman.com; 15 (left), 20-21 (top left), 20-21 (bottom left); Doug Sokell/Visuals Unlimited; 16-17, 25 (bottom right), Taxi/Getty Images; 18 (top left), 35 (middle), Stephen St. John/National Geographic Image Collection; 20-21 (middle), Annie Griffiths Belt/National Geographic Image Collection; 22 (left), Darrell Gulin/Corbis; 22-23 (top), Lockwood, C.C./Animals Animals-Earth Scenes; 22-23 (bottom), J. Eastcott/Y. Eastcott Film/National Geographic Image Collection; 25 (bottom left), Photographer's Choice/Getty Images; 26 (bottom), FoodPix/Getty Images; 28-29 (bottom), Fletcher & Baylis/Photo Researchers, Inc.; 28-29 (top), Claude Nuridsany & Marie Perennou/Photo Researchers, Inc.; 30 (bottom), Comstock Images/Getty Images; 31 (top right), Frank Seguin/Corbis; 31 (middle left), Lynwood M. Chace/Photo Researchers, Inc.; 31 (middle right), Veronique Burger/Photo Researchers, Inc.; 32, 35 (top), Jim Sugar/Corbis.

Produced through the worldwide resources of the National Geographic Society, John M. Fahey, Jr., President and Chief Executive Officer; Gilbert M. Grosvenor, Chairman of the Board; Nina D. Hoffman, Executive Vice President and President, Books and Education Publishing Group.

PREPARED BY NATIONAL GEOGRAPHIC SCHOOL PUBLISHING
Ericka Markman, Senior Vice President and President, Children's Books and Education Publishing Group; Steve Mico, Senior Vice President, Editorial Director, Publisher; Francis Downey, Executive Editor; Richard Easby, Editorial Manager; Bea Jackson, Director of Layout and Design; Jim Hiscott, Design Manager; Cynthia Olson, Art Director; Margaret Sidlosky, Illustrations Director; Matt Wascavage, Manager of Publishing Services; Sean Philpotts, Jane Ponton, Production Managers; Ted Tucker, Production Specialist.

MANUFACTURING AND QUALITY CONTROL
Christopher A. Liedel, Chief Financial Officer; Phillip L. Schlosser, Director; Clifton M. Brown III, Manager

CONSULTANTS AND REVIEWERS
Kefyn M. Catley Ph.D., Assistant Professor of Science Education, Department of Teaching and Learning, Peabody College, Assistant Professor of Biology, Vanderbilt University, Research Associate, Division of Invertebrate Zoology, American Museum of Natural History, New York

Julie Edmonds, Associate Director, Carnegie Academy for Science Education, Carnegie Institution of Washington

◀ Many kinds of plants live in this desert in Uluru National Park in Australia.

Contents

Build Background   **4**
Many Kinds of Plants

1 Understand the Big Idea   **6**
Plants on Earth

2 Take a Closer Look   **16**
Survivors in the Sand

3 Make Connections   **24**

Extend Learning   **30**

Glossary   **34**

Index   **36**

BOOK DEVELOPMENT
Amy Sarver

BOOK DESIGN/PHOTO RESEARCH
3R1 Group, Inc.

Copyright © 2006 National Geographic Society.
All Rights Reserved. Reproduction of the whole or any part of the contents without written permission from the publisher is prohibited. National Geographic, National Geographic School Publishing, National Geographic Reading Expeditions, and the Yellow Border are registered trademarks of the National Geographic Society.

Published by the National Geographic Society
1145 17th Street N.W.
Washington, D.C. 20036-4688

ISBN-13: 978-0-7922-5406-5
ISBN-10: 0-7922-5406-6

2012
    4 5 6 7 8 9 10 11 12 13 14 15

Printed in Canada.

Build Background

Many Kinds of

Plants come in many sizes and shapes. The tallest plants are trees. Some stand higher than tall buildings. Other plants are very small. Some are smaller than your hand.

All plants are important. Why? Without them, we could not **survive**. Plants make food. Plants also make **oxygen**. Oxygen is a gas in the air we breathe. Plants are important to all living things. Look at the photos. How are the plants alike? How are they different?

survive – to stay alive
oxygen – a gas made by plants

Plants

▲ This crocus is a small plant with purple flowers.

▲ These sunflowers have large, yellow flowers.

▲ The redwood tree is one of Earth's largest plants.

1 Understand the Big Idea

Big Idea
Plants have features that allow them to live in many places.

Set Purpose
Learn how plants survive.

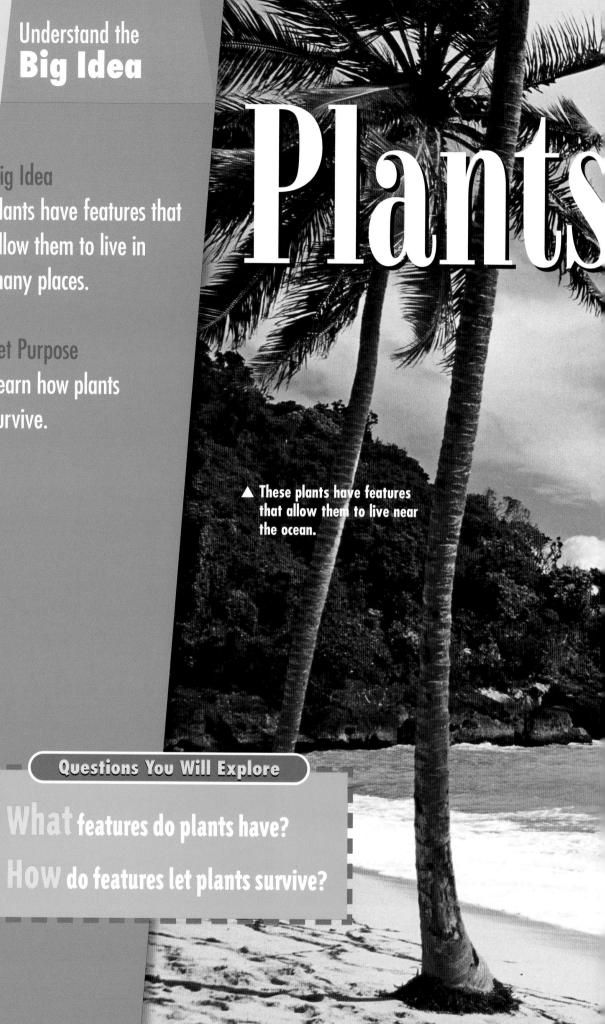

Plants

▲ These plants have features that allow them to live near the ocean.

Questions You Will Explore

What features do plants have?

How do features let plants survive?

on Earth

Plants live in many places. Some live in water. Others live in dry deserts. Plants can be found in city parks and rain forests. How can plants live in so many places?

Each kind of plant has **features** that help it survive. Plants are living things. Plants need air, water, and light to survive. In this book, you will learn about plants and how they survive.

feature – a part of a living thing

Plants in the ocean

Plants in a city park

Plants in a desert

Plants in a rain forest

Leaves
Leaves make food.

Plant Parts

Plants have parts that help them survive. Many plants have leaves, stems, and roots. Each of these parts helps a plant get the things it needs.

Plants need sunlight, water, and air. Plants also need **minerals**. Minerals are materials found in the soil that plants need to live and grow. Leaves, stems, and roots let plants take in minerals, sunlight, water, and air.

mineral – a nonliving material that plants take in through their roots

Roots
Roots grow into the soil. They take in water and minerals from soil. They hold the plant in the ground. They can also store food for a plant.

Stem

A stem holds a plant upright. It carries water and minerals from the roots. It also carries food from the leaves to other parts of the plant. A trunk is the stem of a tree.

▲ These ferns grow on a forest floor.

spores

New Plants

There are many kinds of plants in the world. Plants make new plants in different ways. Some plants grow from tiny structures called **spores**. Plants such as ferns grow from spores.

spore – a tiny structure needed for plants, such as ferns, to make new plants

▲ The sunflower makes seeds that can grow into new plants.

Seed Plants

Some kinds of plants grow from **seeds**. Seeds are plant parts that can grow into new plants. Many of the plants that you see every day grow from seeds. For example, some plants grow cones that make seeds. Other plants need flowers to make seeds.

seed – a plant part that can grow into a new plant

▶ This pinecone has seeds that can grow into a pine tree.

Photosynthesis

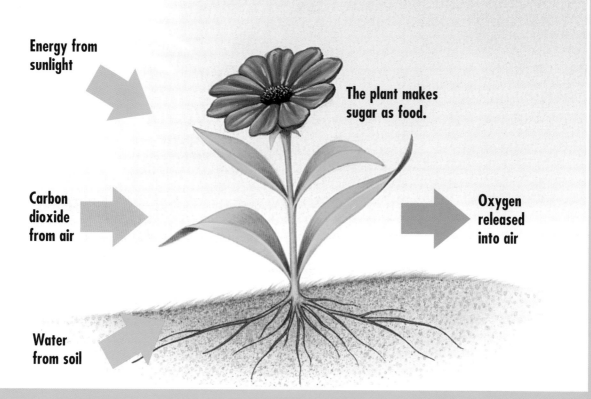

▲ During photosynthesis, plants use sunlight, carbon dioxide, and water to make sugar and oxygen.

Plants Make Food

Seeds and spores are not the only things that plants can make. Plants also make their own food. This is called **photosynthesis.** How do plants make food? They take in water from the soil. Plants also take in **carbon dioxide.** Carbon dioxide is a gas in the air. Plants use the energy of sunlight to change water and carbon dioxide. As a result, oxygen and sugar are made. Plants use this sugar to live and grow.

photosynthesis – the process in which plants use water, carbon dioxide, and sunlight to make sugar and oxygen

carbon dioxide – a gas in the air that plants use to make food

Animals Get Energy From Plants

Plants trap energy from the sun during photosynthesis.

Some animals get energy by eating plants.

Some animals get energy by eating animals that eat plants.

▲ Animals get energy by eating plants or animals that eat plants.

Energy From Plants

Plants get energy from the food they make. Animals also get energy from plants. How? Animals get energy by eating plants. Or animals eat other animals that have eaten plants. The sugar that plants make is used by both plants and animals!

▼ The ground squirrel gets energy from the plants it eats.

Some Places Where Plants Live

Desert

Rain forest

Ocean

Plants in Many Places

Plants are important to all living things. So it is good that plants live in many different environments. Deserts, rain forests, and oceans are just a few of the environments in which plants can live. Plants can be found nearly everywhere on Earth.

▲ The cactus can store water in its thick stem.

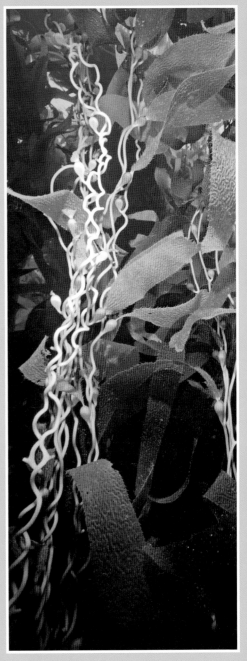
▲ This kelp has long leaves that take in sunlight under the water.

Features Let Plants Survive

Plants that live in different environments may not look the same. That is because each kind of plant has features that help it survive where it lives. Each plant has parts that are suited to its environment. These special features allow plants to live in very different places on Earth.

Stop and Think!

What features help plants survive?

2 Take a Closer Look

Recap
Describe some parts that let plants survive.

Set Purpose
Discover how plants can survive in the hot, dry desert.

▲ Many kinds of plants live in this desert in Arizona.

SURVIVORS IN THE SAND

A desert is a harsh place. During the day, a desert can be very hot. At night, it can be cold. A desert is dry. There is not much water. Yet plants live there. You might be surprised what some desert plants can do.

▲ The prickly pear cactus has spines.

Lose the Leaves

Desert plants have features that help them live where it is very dry. The cactus is one kind of desert plant. A cactus can have sharp **spines**. Spines help desert plants keep the water they take in. Some plants lose a lot of water through their leaves. Spines do not lose as much water. Spines help plants live from one rainfall to the next.

spine – a short, sharp part of a plant or animal

▲ The stems of this cactus can use sunlight to make food.

stem

Stems at Work

The stem of a cactus also helps it survive. The stem moves food and water through the plant. But the stem of a cactus also does something else. It does the same job that leaves do in other plants. The stem makes food. The plant uses water, carbon dioxide, and light. This makes oxygen and the sugar that a cactus uses as food.

The thick stem of a cactus often has a smooth, waxy coating. This helps keep water inside the plant. Some cactus stems even swell up to store more water until the next rain.

stem

▲ The stem of this cactus is swollen to store more water.

Desert Roots

Roots bring water into a plant. But in the desert, water is hard to find. How do plants survive?

Some plants have roots that grow deep into the ground. These roots are more likely to find underground water.

Some roots work the opposite way. Some desert plants can have very **shallow** roots. These roots stay close to the surface. They stretch out from the plant. The roots cover a lot of ground. So they can take in more rain whenever it falls.

Some plants have both kinds of roots. The pinyon pine is a tree with deep roots and shallow roots. Both kinds of roots help the tree. The roots take in as much water as possible.

....
shallow – not deep

Deep Roots
Some desert plants have deep roots that take in underground water.

Shallow Roots
Some desert plants have shallow roots that can take in a lot of rain.

Deep and Shallow Roots
Some desert plants have deep and shallow roots that take in as much water as possible.

Root Systems

◀ The ocotillo plant grows leaves after a rain.

Changes in the Desert

Timing is important for desert plants. Many kinds of plants survive by changing at certain times.

The ocotillo is a tough desert plant. After a rain, the plant grows leaves. Then flowers bloom. Soon, the plant drops its leaves. The plant may look dead. But it is not. It is just waiting for the next rain. These changes help the plant save water. And that helps the ocotillo plant survive in the desert.

▶ Many plants have features that let them survive in the desert.

◀ The ocotillo plant drops its leaves and waits for the next rain.

Surviving in the Desert

Many different kinds of plants live in the desert. Each kind has features that help it survive. Not much rain falls in the desert. So desert plants have roots that let them get as much water as possible. Stems and spines can also do many jobs, such as making food and saving water. Each kind of plant has features that help it live in the dry desert.

Stop and Think!

How do some plants survive in the desert?

3 Make Connections

Recap
Explain some features that allow plants to survive in the desert.

Set Purpose
Read these articles to learn more about plants and how they grow.

CONNECT WHAT YOU HAVE LEARNED

A World of Plants

Plants live almost everywhere. Each kind of plant has parts that allow it to survive.

Here are some ideas you learned about plants.

- Many plants have roots, stems, and leaves.
- Some kinds of plants grow from spores and some grow from seeds.
- Plants make food during photosynthesis.
- Plants have features that help them survive in different places.

Check What You Have Learned

What do the pictures show about how plants survive?

▲ The leaves, stem, and roots on this tree help it survive.

▲ The sunflower has seeds that can grow into new plants.

▲ Plants use sunlight, water, and carbon dioxide to make food and oxygen.

▲ Spines help this cactus save water in the dry desert.

CONNECT TO FOOD

Plant Foods

▲ Broccoli is a vegetable that can protect you from disease.

Have you ever been told to eat your fruits and vegetables? The fruits and vegetables you eat are plants or plant parts. These plants help you stay healthy. They give you vitamins and minerals that your body needs. Eating fruits and vegetables, such as blueberries and broccoli, can protect you against disease.

▲ Blueberries

CONNECT TO MEDICINE

Pills From Plants

Many plants help cure human illnesses. Native Americans have known this for a long time. They have used plants to treat everything from stomach pains to toothaches.

Today, scientists still use plants to make medicines. They also search the world to discover new plants. New plants might help people in new ways.

▼ For many years, people have used purple coneflowers to make medicine.

CONNECT TO ANIMALS

Plants That Eat Meat

Did you know that some plants eat animals? It may sound strange. But it is true. These plants are carnivorous. That means they eat meat.

Most carnivorous plants are not as scary as they sound. They usually just trap tiny insects. Then they break down the insects' bodies for food. Most carnivorous plants also make their own food.

▲ This plant traps insects for food.

CONNECT TO REPRODUCTION

A Plant's Smell

The world's largest flower is also one of the smelliest on Earth. The *Rafflesia* flower is about three feet wide. But it would not make a very good gift. Why? This huge flower smells like rotting meat. The smell attracts flies. The flies help the *Rafflesia* plant make seeds.

◀ The *Rafflesia* flower smells like rotting meat.

Extend Learning

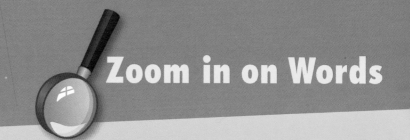

Zoom in on Words

Many kinds of words are used in this book. Here you will learn about compound words. You will also learn about multiple-meaning words.

Compound Words

Compound words are made by joining two shorter words. Find the compound words below. What smaller words form each compound word?

under + ground = underground

The deep roots can take in **underground** water.

sun + light = sunlight

Sunlight fills the sky.

Multiple-Meaning Words

A multiple-meaning word is a word that has more than one meaning. Find the multiple-meaning words below. What two meanings do each of these words have?

The **soil** is dark brown.

They **soil** their clothes with mud.

Roots can **store** food for a plant.

She buys food in a **store.**

The **leaves** are green.

He **leaves** the cab.

Research and Write

Write About a Plant

Research plants on your own. Find out what plants need to grow. Write journal entries telling what you learned.

Research

Get a pack of seeds, soil, and three plastic cups. Put the soil in the cups. Then plant the seeds. Number the cups. Give one cup water and sunlight. Give another cup water only. Give the last cup sunlight only.

Watch and Take Notes

Look at what happens to the plants in each cup for two weeks. Make a chart. Use the chart to take notes and draw pictures of what you see in each cup.

Write

Keep a journal to write down what you see each day. After two weeks, write a paragraph explaining what plants need to grow.

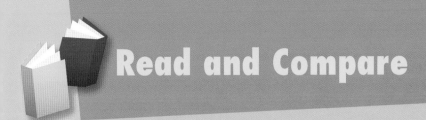

Read and Compare

Read More About Plants

Find and read other books about plants. As you read, think about these questions.

- What do plants need to survive?
- What features help plants survive in different places?
- How do scientists learn more about plants?

Books to Read

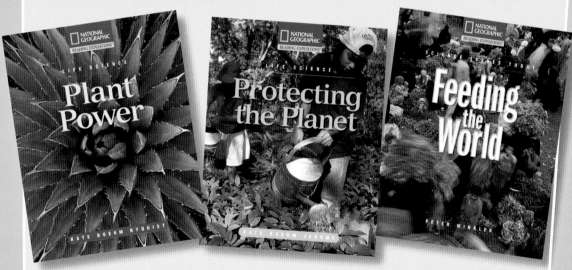

▲ Read about plants and why they are important.

▲ Read about plants and other resources on Earth.

▲ Read about plants and world hunger.

Glossary

carbon dioxide (page 12)
A gas in the air that plants use to make food
During photosynthesis, plants use carbon dioxide from the air.

feature (page 7)
A part of a living thing
The purple flower is a feature of this plant.

mineral (page 8)
A nonliving material that plants take in through their roots
Roots take in minerals from the soil.

oxygen (page 4)
A gas made by plants
Living things need the oxygen plants release into the air.

photosynthesis (page 12)
The process in which plants use water, carbon dioxide, and sunlight to make sugar and oxygen
Plants make their own food during photosynthesis.

seed (page 11)
A plant part that can grow into a new plant
A seed can grow into a new plant.

shallow (page 20)
Not deep
This plant has shallow roots.

spine (page 18)
A short, sharp part of a plant or animal
This cactus spine is sharp.

spore (page 10)
A tiny structure needed for plants, such as ferns, to make new plants
A fern grows from a spore.

survive (page 4)
To stay alive
This cactus has features that let it survive in the desert.

Index

air	4, 7–8, 12, 34
carbon dioxide	12, 19, 34
carnivorous plants	28
desert	3, 7, 14, 16–25
feature	6–7, 15, 18, 22–24, 33–34
leaves	8, 18, 22–25, 31
mineral	8–9, 26, 34
oxygen	4, 12, 19, 34
photosynthesis	12–13, 24–25, 34
roots	8–9, 20–21, 23–25, 31, 34–35
seed	11, 32, 35
shallow	20–21, 35
spines	18, 23, 25
stems	8–9, 15, 18–19, 23–25
sunlight	8, 12, 18–19, 25, 30, 32, 34
survive	6–8, 15–16, 19–20, 22–25, 33, 35